Ernst G. Siebeneicher-Hellwig and Jürgen Rosinski

Forging Damascus Steel
Knives for Beginners

4880 Lower Valley Road • Atglen, PA 19310

Other Schiffer Books on Related Subjects:

Making Integral Knives, 978-0-7643-4011-6, $24.99

Making Leather Knife Sheaths Volume 1, 978-0-7643-4015-4, $24.99

Pocketknife Making for Beginners, 978-0-7643-3847-2, $29.99

The Lockback Folding Knife: From Design to Completion, 978-0-7643-3509-9, $29.99

Basic Knifemaking: From Raw Steel to a Finished Stub Tang Knife, 978-0-7643-3508-2, $29.99

Originally published as *Damast-Messer schmieden für Anfänger* by Wieland Verlag GmbH.
Cover design by Bruce Waters
Layout by Caroline Wydeau
Translated from German by Ingrid Elser and John Guess

Published by Schiffer Publishing, Ltd.
4880 Lower Valley Road
Atglen, PA 19310
Phone: (610) 593-1777; Fax: (610) 593-2002
E-mail: Info@schifferbooks.com

For the largest selection of fine reference books on this and related subjects, please visit our website at

www.schifferbooks.com. You may also write for a free catalog.

This book may be purchased from the publisher.
Please try your bookstore first.

We are always looking for people to write books on new and related subjects. If you have an idea for a book, please contact us at

proposals@schifferbooks.com

Schiffer Books are available at special discounts for bulk purchases for sales promotions or premiums. Special editions, including personalized covers, corporate imprints, and excerpts can be created in large quantities for special needs. For more information contact the publisher.

In Europe, Schiffer books are distributed by
Bushwood Books
6 Marksbury Ave.
Kew Gardens
Surrey TW9 4JF England
Phone: 44 (0) 20 8392 8585; Fax: 44 (0) 20 8392 9876
E-mail: info@bushwoodbooks.co.uk
Website: www.bushwoodbooks.co.uk

Type set in Courier10 BT/Zurich BT

ISBN: 978-0-7643-4012-3
Printed in China

ACKNOWLEDGMENTS

A special thank you to Mrs. Petra Steinberger, manager of the company Dick – Feine Werkzeuge, for her support. Manfred Ritzer, Herbert Meyer, Klaus Wiesent, Hans-Georg Ortner, Günther Schmid, and Hubert Ziegler provided invaluable help creating the images used in this volume. For this book, Manfred Ritzer also developed a compact and very powerful gas forge. Dr. Michael Maria Rind, the official archaeologist of the county Kelheim at that time, provided images of the bloomery project at the Archaeology Park in Altmühltal, Germany.

CONTENTS

PREFACE

The great success of *Basic Knifemaking: From Raw Steel to a Finished Stub Tang Knife* in this workshop series and the positive response from our readers has encouraged us to write a second book which builds on the first one and deals with forging Damascus steel. The brick forge described in the first book has proven itself worthy. Meanwhile, ten courses with ten participants each were held at the workshop of the company Dick in Metten, Germany, around the theme "Simple Forging – Simply Forging." The basis for these courses was the concept of our brick forge.

A FEW WORDS UP FRONT

As a hobby, forging has been booming in recent years. More and more people are discovering how much joy it can be to shape a piece of steel with fire and hammer—an archaic craft which holds a lot of fascination, especially in our modern times.

Forging Damascus steel, or just Damascus, is something like a royal discipline among steel forgers. To weld several layers of steel together in the fire, and to turn it into a single, coherent piece requires some amount of skill and experience. But it is no witchcraft. You don't even need much to do it: a forge, a hammer, an anvil.

With this volume we show you how to build a gas or coal forge yourself with little effort. The hammer can be bought at the hardware store; as an anvil you can use a piece of railroad track. Thus prepared you can start immediately!

Ernst G. Siebeneicher-Hellwig and Jürgen Rosinski show you what can be done with this cheap equipment. They forge Damascus, create various patterns, forge a blade, and construct a complete knife.

The Knife Magazine (MESSER MAGAZIN) workshop series assembles a multitude of themes all around knifemaking in a way which enables you not only to follow each step but to do it yourself. We emphasize the practicality of all the volumes for use in the workshop.

Thus all the volumes are provided with a wire binding. This way the book stays open when you put it down on the workbench. Also, we made sure the images and text are large enough to see when the book is lying next to you during work.

We have tried to explain every step of work in the most comprehensive way. But before you pick up your tools, you should read all the descriptions and explanations in this book first. This way you'll know what to expect and won't be confronted with unpleasant surprises later on.

As you use this book I hope your work is fun and successful!

Hans Joachim Wieland
Chief Editor, *Knife Magazine* (*MESSER MAGAZIN*)

A BRIEF HISTORY OF DAMASCUS STEEL

Since the beginnings of shaping steels, Damascus steel has created a perpetual fascination. The letter of appreciation for the *"wyrmfāh"* (decorated with worms, snakes, or dragons) blades Theoderic the Great received as a gift from the king of the Lombards speaks of the high value of Damascus blades in the late classic period. The Damascus swords of the Saracens are said to have cleanly cut through the blades of the Christian knights in battles during the crusades. Besides its fascinating aesthetic, the steel made of many layers also has its practical advantages, even now during the age of high-alloy tool steels, as we will show during our discussion on Japanese kitchen knives (page 21).

1.1. The Development of Damascus

It was not until much later than bronze, the first alloy suitable for producing tools and weapons, that iron was used. The reason was that our ancestors didn't know how to create the heat necessary for producing steel. Only after inventing the bloomery furnace was it possible to produce steel from iron ore in a charcoal fire.

Prior to that, another technical innovation was necessary: charcoal. If wood is heated to glowing temperature without the presence of oxygen, the wood doesn't burn up. Instead, a superb material for burning is created, charcoal. Only in modern times has the discovery of how to change hard coal into coke allowed the production of steel on an industrial scale.

About 3,700 years ago, the Hittites started smelting iron ore. The legitimate question arises: What caused the early smiths to heat inconspicuous looking stones to the greatest heat which could be created at that time for many hours? The most plausible explanation is that the beginnings of metallurgy are rooted in the production of ceramics. Here temperatures of similar intensity are created, and for producing colorful glaze metalliferous minerals were used.

At the beginning of iron metallurgy there were two kinds of iron: soft wrought iron and brittle cast iron. These qualities were determined by the content of carbon. Minimal carbon content (less than 0.4%) leads to steel which can't be hardened. A very high percentage of carbon makes a very hard, but brittle steel which breaks easily. For swords and knives, the main tools benefiting from steel technology, both kinds were of only limited availability. Soft swords bent, and hard blades broke from strong blows.

Many centuries ago, probably in present-day Anatolia (eastern Turkey), some clever people hit on the idea to combine the qualities of both kinds of steel by welding them together in the forging fire. The result was a sword which was flexible enough for strong blows, but also cut well. The owner of such a sword was vastly superior to his opponent. The secret of creating Damascus steel was well guarded, for it was often decisive in war. The "magical swords" were surrounded by a supernatural aura which was enhanced by their unusual optics. The bearer of such a sword did not only benefit from its better quality, he also had the power of myth on his side which preceded these swords and created a psychological advantage in a fight. It is no wonder that these Damascus swords cost a fortune and thus were only privy to relatively few people.

But how did humanity discover Damascus? In the bloomery, also called a Renn furnace, pieces of iron came into being, some of which were moldable (with low carbon content), while others with high carbon content broke like ceramics. To create a knife, an axe, a plough, or a sword, these small pieces had to be welded together. While doing so, the early smiths realized that some of their products were harder than others but didn't break under stress. What was chance at the beginning turned into intention later on. The bladesmiths deliberately used hard steel for the cutting edge of the tools and left the rest of the area soft and thus more flexible. Now the tool had a hard cutting edge without breaking during use.

BLOOMERY FURNACE PROJECT

In September 2008, iron ore was smelted during a bloomery furnace project at the Celtic Forge of the Archaeological Park in Altmühltal, Germany. In the bloomery, iron is not melted. Instead, the iron ore—a mixture of iron and rocks not containing any iron—is reduced. In chemistry "reducing" means the separation of chemical compounds. The iron within the iron ore is mainly there in the form of iron oxide, a bond of iron and oxygen. In the heat of the bloomery furnace, the oxygen reacts with the carbon of the charcoal and iron is left.

The product of this process is called bloom. It is a conglomerate of iron, residues of iron ore which don't react, slag, and unburnt charcoal. With this substance you don't have usable iron yet. To get rid off the unwanted impurities, several cycles of heating and forging are necessary.

The bloomery furnace in the valley of the river Altmühl was charged with ore from Erzberg close to the Austrian town Eisenerz. The iron ore of this area, called Ferrum Noricum, was praised for its quality in Roman times.

For this project 18 sacks of charcoal and 36 kilograms (~80 lbs) of iron ore were used. The recovered bloom had a weight of ten kilograms (22 lbs). The bloomery was in operation for 24 hours. You can see that this process is very material and labor intensive, especially if you add time for building the furnace, searching for iron ore, and burning the charcoal. In Europe, by the way, in ancient times whole areas were deforested for producing charcoal.

That it was possible for the early smiths to produce iron by means of the bloomery is one of the most important pioneering feats in the history of humankind. It is no wonder that humans who could produce iron from stone in those times were said to have magical abilities.

Photo: Prof. Dr. Rind

The bloomery furnace in front of the Celtic Forge at the Archaeology Park in Altmühltal, Germany.

Photo: Prof. Dr. Rind

The bottom of the furnace is made of clay, the structure is reinforced with willows.

The bloomery furnace is fueled with charcoal and iron ore.

Piece of charcoal from Styria, Austria.

Bean ore from the Swabian Jura, Germany.

Bog iron ore from Altmühltal.

During the "tapping," the bloomery furnace is opened at its side.

The opened bloomery furnace after tapping.

The result of the reduction process is the bloom.

Here the bloom is cut apart and polished.

WOOTZ – DAMASCUS OUT OF THE CRUCIBLE

Wootz is a kind of steel which is produced in the melting pot. Different from welded Damascus, in a special melting process a steel is created which has the best qualities for the production of swords. What is achieved for welded Damascus by welding together soft and hard steels, is achieved for wootz during the process of melting: Areas with high carbon content and others with low carbon content are created.

Since these areas inside the steel get well distributed during the process of forging, a final product is achieved which meets the requirements: a sword with a flexible blade which is also resistant to wear.

The raw materials for wootz and bloom from the bloomery furnace.

A clay crucible with the bloom.

Two crucibles filled with bloom, leaves, and glass. The leaves cause the carbonization of the steel.

The filled crucible is heated in the gas forge to melting temperature.

The glass melts and creates a layer impermeable to air. Thus the carbon inside the bloom can't burn up.

The crucible is broken and the "wootz-cake" is released.

The inside of the crucible: it is clearly visible how the mixture melted down.

1.2. Regional Developments

Swords of the Germanic Merovingians and Alamanni were produced with a soft core and a hard cutting area—the cutting edge was partially hardened using a selective process. The Vikings in the north built their swords of twisted Damascus bars with cutting areas of hard steel. In the Indo-Persian area a totally different technique was developed with wootz: In an ingenious melting process, adding different substances (e. g. charcoal, leaves, or other organic materials) created alternating soft and hard zones inside the steel structure.

The Kris dagger in Indonesia, in comparison, was produced by welding. Material was also used which was said to be of divine origin because it fell down from the sky in the form of meteorites. The Kris was said to have magical powers, good as well as evil ones.

In the Far East, in Japan, the art of forging swords was brought to its pinnacle. The smiths of the samurai produced swords with Damascus blades which were folded many times and possessed a very hard and very sharp cutting edge which was hardened selectively. That these katanas were in part tested on prisoners is proven by inscriptions on the tangs of some of these swords. It is also reported that glowing swords were pushed into the bodies of people sentenced to death and that the swords were thus hardened by the victims' blood.

Damascus steel, short: Damascus, wasn't invented in the city of Damascus, as you might think because of its name. On the contrary, Damascus was a center of commerce during the Middle Ages. And many weapons were traded there as well.

Nowadays, many very efficient mono tool steels are available. So there is no need anymore to produce Damascus steels for technical reasons. But the outstanding aesthetics still exist: Damascus is simply beautiful. Besides that, a knife made from Damascus is of higher value because more time and effort has been invested in making it.

JAPANESE KITCHEN KNIVES

There is at least one exception for which multi-layered steel still makes technical sense and is even necessary: traditional Japanese kitchen knives. Like samurai swords, these knives are made of several layers. Soft iron and hard steel are welded to create a usable knife with a cutting edge of legendary sharpness. Such extremely sharp edges can—for technical reasons described in more detail later on—only be created from carbon steel of extreme purity. This kind of steel has to be protected by layers of softer steel to prevent it from breaking.

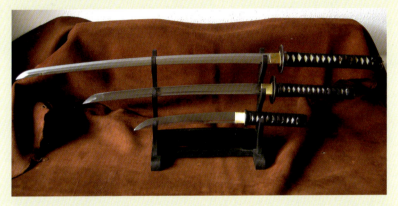

Katana (sword), wakizashi (short sword), and tanto (dagger) with hamon (hardening line).

Hamon: This is the borderline between hard and soft areas of the steel.

When in 1876, in the course of a government reform, the samurai caste was dissolved and wearing and producing samurai swords was prohibited, the skills to create such blades wasn't lost. A few blades-miths took to the production of kitchen knives and used the traditional techniques on them. They welded the steel with high carbon content to softer steel and thus continued the centuries-old tradition. This is one of the reasons why Japanese kitchen knives have a good reputation and rightly so.

For the production of the knife steel, iron sand is used which is pulled out of the river Hino by means of magnets. In earlier times, the steel was melted in a Renn furnace (bloomery furnace). Even today this technique is still sometimes used. The product of the Renn process is called *Tamahagane* and is a mixture of slag and iron. The relatively small amount of the sword steel produced at such great cost is not released on the market but provided to chosen swordsmiths. Today a few mastersmiths again have the license to produce swords.

For the cutting area of traditional Japanese knives, two kinds of carbon steel are used: white paper steel (*Shiro Gami*) and blue paper steel (*Ao Gami*). Both are carbon steels of high purity with small amounts of chromium and tungsten alloyed to the blue paper steel. Thus this steel becomes a bit more tough, but the grain isn't as fine as with the white paper steel. This means the steel can't be ground as finely. So for

Tamahagane raw material.

Photo: Dick Feine Werkzeuge

knives for which the achievable degree of sharpness is of utmost importance (for example, kitchen knives) white paper steel is preferred. If the knife needs to be tougher, as is the case for hunting knives, the smith prefers to use the blue paper steel. The names, by the way, originate from the paper in which the steel is wrapped. To avoid mixing them up, each quality is wrapped in paper of different color.

For Japanese cooks, food preparation is a kind of cult. For this, a good knife is very important. A master cook in Japan is honored with the name *Hocho Jin* which means "man of the knife." This shows that knives are held in high esteem. Japanese kitchen knives, rightly so, have a good reputation for being extremely sharp. This sharpness can only be achieved on an operational level by means of multi-layered steel. Because of this sharpness, the food isn't squashed but cleanly cut. More flavor stays inside the food because fewer cells are damaged by the clean cut.

PARTS OF A JAPANESE BLADE BEFORE WELDING

Multi-layered steel, not hardenable

White paper steel, hardenable

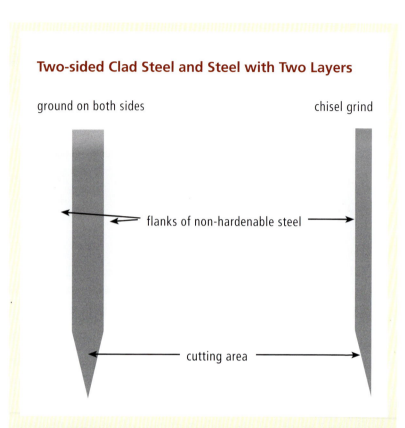

Two-sided Clad Steel and Steel with Two Layers

ground on both sides   chisel grind

flanks of non-hardenable steel

cutting area

**Japanese kitchen knife
with Suminagashi blade
(Damascus).**
Photo: Dick Feine Werkzeuge

**A Japanese vegetable knife
(Usuba) with three layers,
made from a Japanese
officer's sword after Japan's
capitulation in World War II.**

EQUIPMENT AND MATERIALS

Gas or charcoal? What ought to be used to produce the heat necessary for forging? When deciding on a forge, the reader has to face this basic question. Charcoal is easy to get but produces smoke and ashes. A charcoal fire builds up the carbon content in the steel while gas reduces it (slightly). Gas is clean and doesn't create much of a smell. We have to leave the decision on how to generate heat to the reader, since both variants lead to good results.

2.1 Gas Forge

There are many possibilities to choose from for building a gas forge yourself. Here we want to introduce a gas forge, built from a flue tube and designed by knifemaker Manfred Ritzer, for forging knife blades and other small items. Building it is relatively easy. The forge works very efficiently, nevertheless. It heats up quickly, uses gas efficiently (gas pressure while welding is about 2.0 bar [29 psi], while forging about 0.5 bar [7.25 psi]), and creates high temperatures.

Manfred Ritzer with the gas forge he developed.

The gas forge in its nascent stage. **First test run.**

It works! The finished forge in action.

PARTS LIST FOR THE FLUE TUBE FORGE

No.	Designation	Measurements	Price ($)
1	reducing pipe joint	1 1/4 x 1"	4.70
2	mixing tube	Ø 33.5 x 3.0 x 130.0 mm	2.70
3	flue tube, reduced	Ø 150.0 x 250.0 mm	23.00
4	gas nozzle	0.6 mm	1.35
5	double nipple	2 x G3/8L	5.75
6	burner pipe	Ø 27.0 x 2.5 x 210.0 mm	2.70
7	flat iron	70.0 x 170.0 x 4.0 mm	2.70
8	YTONG building slab	599.0 x 199.0 x 50.0 mm	1.35
9	cover brick	240.0 x 115.0 x 71.0 mm	1.35
10	flue socket with 1" thread	100.0 mm	4.00
11	gas feed	brass rod Ø 35.0 mm	4.00

The parts of the forge: flue tube, reducing pipe-joint, flue socket, and two flat pieces of iron for support.

The insulation of the flue tube is made using a fireproof ceramic mat which is glued in by means of fireproof stone putty and then plastered to achieve a smooth surface. While working with fibrous material you should always wear a mask or respirator!

After lining the combustion area, plan to dry the parts for a few days. The first heating has to be done very carefully to prevent cracks in the lining.

The finished forge casing with the socket for the burner. The socket is at an angle of about 45° and is located a bit to the back of the casing.

View of the forge from its side.

A view of the forge standing on end.

The burner.

Burner pipe with end stop.

The mixing tube. The oblong hole was milled. But it can also be created by drilling several holes and filing the shape out.

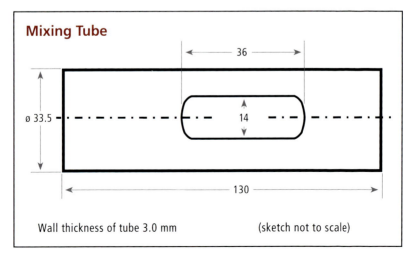

Mixing Tube

36

ø 33.5

14

130

Wall thickness of tube 3.0 mm   (sketch not to scale)

The gas feed with gas nozzle.

Gas Feed

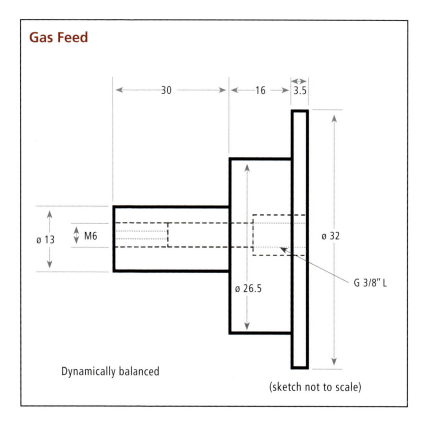

30   16   3.5

ø 13   M6   ø 32

ø 26.5   G 3/8" L

Dynamically balanced

(sketch not to scale)

The Complete Burner

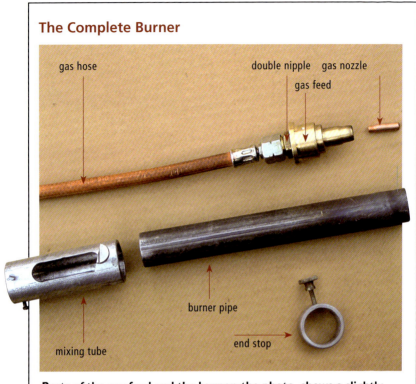

gas hose

double nipple   gas nozzle

gas feed

burner pipe

end stop

mixing tube

Parts of the gas feed and the burner: the photo shows a slightly more complex version, but the drawn version works as well.

FUNCTION OF THE VARIOUS BURNER PARTS

double nipple	connection between gas hose and gas feed
gas feed	feeds the gas to the gas nozzle
gas nozzle	swirls the gas
mixing tube	mixes gas and air
adjusting screw	fixes the burner pipe inside the mixing tube
end stop	fixes the burner inside the casing of the forge

The burner is pushed inside the flue socket on the casing and the gas is turned on at the propane bottle. Ignition is best done by means of a piece of lit newspaper.

The flame can be regulated by changing the air inlet at the mixing tube. This can be done by varying how far the burner pipe is pushed into the mixing tube. When the position is optimal, fix it with the adjusting screw.

After its use, the burner should be taken out of the kiln to avoid heating it up unnecessarily. In any case, there should be protection against rupture of the hose, and the forge should never be unattended while it is running!

We hope that every reader who is somewhat skilled with building things can reproduce a burner and forge with the help of the parts list and the photos.

WARNING: If you do not have experience working with gas, these steps are risky! Observe the relevant safety regulations for working with gas. The authors and the publisher can not be held responsible for personal or material damage caused by building or using the described gas forges.

A GAS FORGE MADE OUT OF AN AMMUNITION CRATE AND A GAS BOTTLE

This method was seen at a knife show in Olching, Germany. The forge was constructed by the competent and hard-working people at the Kursschmiede workshop.

Ammunition crate forge with burner.

The burner is placed off-center and vertically. By doing this, the flames circulate in the combustion area and the heat is distributed more evenly.

The forge is heated up.

View from the side.

GAS FORGE FROM A GAS BOTTLE

An empty propane bottle is also suitable for the construction of a gas forge.

The casing of the forge with welded-on handles and burner socket. Here, the burner is also placed at the side and off-center to let the flames circulate. It is easy to see that the gas tank was cut apart in the center and welded together again when the insulation lining was installed.

The forge with the burner and the gas feed.

2.2 Brick Forge

In the workshop volume, Basic Knifemaking: From Raw Steel to a Finished Stub Tang Knife, we already described in detail the construction of our brick forge. Thus we only mention this topic briefly here. The basic idea is to build a simple, functional forge with little effort and space. For building this forge, no specialized knowledge is needed.

The forge from the first volume was only modified insofar as an additional cover was mounted to better keep the heat. In addition, we also added a regulator to better regulate the stream of air. Besides that, instead of a hair dryer we used the motor of a vacuum cleaner as a blower.

Fuel for the brick forge is again charcoal. Tell skeptics that it wasn't until modern times that anything other than charcoal was used for forging and Damascus was produced with charcoal 2,000 years ago. Charcoal was not replaced as a fuel source until 1820 with the invention of carbonizing hard coal.

The base of our simple brick forge.

The air pipe with its end cap.

The tube is led through the brick at the front.

The forge without its roof.

Blower: the motor of a vacuum cleaner with a piece of inner tube from a bicycle and hose clamps.

A valve for simple air regulation.

The completed forge.

The charcoal fire is hot enough for forging Damascus.

2.3 Tools for Forging

A decent forging hammer can be constructed out of a reasonably priced hammer from any home improvement store. The same goes, in principle, for the forging tongs. We modified standard tongs to get forging tongs.

Beyond this, there is a multitude of special ready-to-use hammers available—some with specific shapes are in demand in Scandinavia and Japan.

You don't have to buy a real anvil right away: a piece of railroad track is absolutely sufficient at the beginning.

The hammer from the home improvement store (left) and the modified hammer.

Comparing the two hammer faces.

A small selection of hammer types (left to right): Nordic type, Japanese hammer, and three German hammers.

A special hammer (extending in length) that a blacksmith uses for beating out steel together with a striker.

A traditional Japanese forging hammer.

A small but fully functional anvil made out of a piece of railroad track.

Tongs from the home improvement store (above) and handmade forging tongs.

2.4 Types of Steel

For creating welded Damascus in an open fire, only pure carbon steels (like C45, C50, C60, C75, C80, C90, or C100) or low alloy tool steels (see examples in the box below) are suitable. At least one of the steels to be welded must have a carbon content of more than 0.4% in order for the Damascus to be hardened. If not all of the steels in a welded package are hardenable, then at least the steel used later in the area of the cutting edge has to be hardenable.

FREQUENTLY USED TYPES OF STEEL

Material #	Steel Type	Elements of the Alloy in %						
		C	Cr	Mn	V	W	Ni	Mo
1.2510	cold work steel	1.0	0.6	1.0	0.1	0.7		
1.2842	cold work steel	0.9	0.3	2.0	0.1			
1.2767	cold work steel and hot-work steel	0.5	1.5	0.4			4.0	0.3
1.2552	cold work steel	0.8	1.1	0.3	0.3	2.0		
1.3505	cold work steel	1.0	1.5	0.4		2.0		

The percentages are average values.

Because of their high content of chromium, stainless steels can't be welded together without problems. In combination with oxygen, chromium reacts to form chromium oxide on the surface of the steel, and thus prevents welding. For this reason, stainless Damascus is produced in a vacuum or controlled atmosphere, which means it takes a lot of effort, which also has a profound influence on the price.

But even "stainless" blades are not really stainless. It is more realistic to say they are not prone to rust. This better describes the actual fact: Stainless types of steel (from a chromium content of 13% upwards) can rust indeed. Nevertheless, they have a certain rust-resisting property which is sufficient under normal conditions. Real rust-resistant quality is only achieved by non-hardenable stainless steels like V4A or AISI 416.

Since stainless steels suited for making cutlery were developed during the last century, there is an ongoing debate about the pros and cons of stainless and rusting kinds of steel. In the table further down this page, the advantages and disadvantages for both kinds are listed in a clearly arranged way.

The image on the top of page 44 shows the cross section of a non-stainless blade of carbon steel (left) and a stainless blade of alloyed chrome steel (right). The chromium carbides to the right are relatively large, which prevents grinding the cutting edge to a fine blade.

COMPARISON OF QUALITIES: STAINLESS STEELS/RUSTING STEELS

	Sharpness	Re-Sharpening	Appearance	Efforts for Maintenance
Stainless Steels (high-alloy)	average	takes some effort	good	low
Non-Stainless Steels (non-alloy or low-alloy carbon steels)	very good	easy	mottled	high

Because of its inner structure, stainless steel can't be ground as finely as carbon steel. For stainless steel, chromium is alloyed to ensure corrosion resistance. Chromium forms big carbides which prevent extremely fine sharpened cutting edges. By means of their hardness, these carbides enhance the durability of stainless steel blades. On the other hand, re-sharpening the blades becomes more difficult. With carbon steels the appearance suffers, because the steel becomes gray and mottled when it comes into contact with acids which are contained in food. Remedy: use a rust eraser.

Printed courtesy of Dick Feine Werkzeuge GmbHV

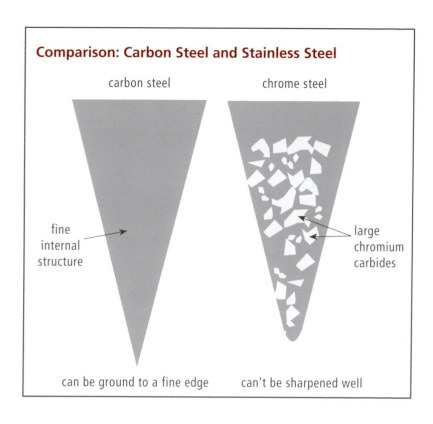

Comparison: Carbon Steel and Stainless Steel

carbon steel

chrome steel

fine internal structure

large chromium carbides

can be ground to a fine edge

can't be sharpened well

A Nordic-style knife by Thomas Steindl with a blade of nickel Damascus.

THREE TYPES OF STEEL IN DETAIL

In the following table we describe one of the tool steels that is most often used in the West (1.2842, also known as O2) and the Japanese "paper steels" in more detail. White and blue paper steels are famous for their especially high purity. Annoying companions of steel like phosphorus and sulfur are only present in very small amounts. Thus Shiro Gami and Ao Gami are especially suited for the production of Damascus steel.

	Material No. 1.2842 (O2)	White Paper Steel (Shiro Gami)	Blue Paper Steel (Ao Gami)
Composition	C 0.9% Si 0.2% Mn 2.0% Cr 0.3% V 0.1%	C 1.1%	C 1.1% Si 0.1% Mn 0.2% Cr 0.3% W 1.0%
Forging	850–1,050°C (1,562–1,922°F)	850–1,050°C	850–1,050°C
Hardening	790–820°C (1,454–1,508°F)	white paper steel 780–800°C (1,436–1,472°F) blue paper steel 820–830°C (1,508–1,526°F) holding time for 6 mm (~0.25 in) material thickness 7 minutes, for 3 mm (~0.125 in) thickness 3 minutes	
Quenching	in oil (hardness after quenching about 64 HRC)	in oil	
Annealing	temp. HRC °C (°F) 100 (212) 63 200 (392) 61 250 (482) 59 300 (572) 56	wait for 10 minutes, then anneal at 150–200°C (302–392°F) (vegetable knife 150°C [302°F], whittling knife 175°C [347°F], chopping knife 200°C [392°F]) time for annealing at least 30 minutes, hardness: chopping knife 62 HRC, whittling knife 63 HRC	

Heat colors for forging and annealing can be found in the appendix.

A special role is played by the non-ferrous metal nickel. Nickel is well suited for creating patterns in Damascus steel, because it can be forged easily and keeps its silvery shine after etching. This can be seen very well on the Damascus blade made by Thomas Steindl for the knife on the bottom of page 44.

ALLOY ELEMENTS AND THEIR EFFECTS

C	carbon	from 0.4% onwards steel becomes hardenable
Cr	chromium	protects against corrosion from 13% upwards, enhances hardenability
Si	silicon	enhances hardness
Mn	manganese	enhances hardness and hardenability
W	tungsten	forms fine carbides, enhances edge-holding
V	vanadium	forms hard carbides and makes the grain finer
Ni	nickel	enhances hardness
Mo	molybdenum	forms hard carbides and makes the grain finer, enhances hardness
Co	cobalt	enhances wear resistance
P	phosphorus	increases brittleness and propensity for breaking
S	sulfur	increases brittleness and propensity for breaking

FORGING DAMASCUS

3.1 Assembling the Package

For our package we stack different steels on top of each other. The pieces have to be free from fat and stains to ensure a clean weld. While using non-hardenable and hardenable steels together, you have to take care that the hardenable steel is placed in such a way that it can be used as the cutting area later on (usually in the middle of the package).

If multiple folding is planned, this also has to be taken into account. Since the layers become thinner through multiple folding and forging out, it is difficult to hit on a hard layer during the making of the blade. With multi-layered Damascus the layers become so thin that there is also soft steel in the area of the cutting edge, which doesn't exactly enhance the quality of the cutting edge. Thus it is recommended to only use hardenable steels for multiple folding.

For knives with Damascus which is rich in contrast (often with a portion of pure nickel or St37), a cutting bar of hardenable steel (or Damascus made from several hardenable steels) is soled on. More on this in the chapter about multiple bar Damascus.

For our Damascus package we used the low-alloy tool steel 1.2510 (100MnCrW4, also known as O1) and the structural steel St37 with a carbon content of about 0.2%. The advantage of this combination is that it leads to a Damascus which is very rich in contrast.

Layers of structural steel and 1.2510 (O1) tool steel.

The package
is welded at
the sides with
electrical
welding
equipment,
heated in the
forge, then
coated with
borax. Ordinary
wire can be
used to fix the
package in place.

For better
handling of
the packages,
welding
on rods is
recommended.

Preparations for a package with lateral layers of laminated steels and a central cutting layer of white paper steel. *Photo: Klaus Wisent*

A package of St37, laminated two-sided clad steel, and white paper steel.
Photo: Klaus Wisent

3.2 Welding the Package

The welding temperature of steel depends on its carbon content. The higher the content, the lower the temperature. If the steel is heated too much, the carbon inside the steel burns up, the steel will start to emit sparks, and the material becomes useless.

If different types of steel are welded together in the forging fire, the temperature must be suitable for all of the types being used. This leads to a very small common range of temperatures. In our case, the needed welding temperature is about 1,200°C (2,192°F), which is quite high. The heat color is light yellow.

First, the package is evenly heated to a bright red hot color (about 900–950°C [1,652–1,742°F]), then it is taken out of the fire again. We add flux (borax) to the sides of the layers. The flux melts in the fire and is sucked between the layers by capillary action. This prevents the creation of slag which can be dangerous to weld. The Japanese smiths, by the way, use a kind of flux powder that also contains boric acid and finely pounded cinder in equal parts.

Then the package is heated to forging temperature. When the package has reached the necessary temperature (yellow-red in color), it is taken out of the fire and welded rapidly with consistent hammer blows starting from one end. Here it is important to drive out the flux from the center of the package to the sides. It should not be locked inside and create bubbles because this would prevent the layers from connecting.

"Packing" the package prior to the actual welding is recommended. During this process it becomes compressed, so that unevenness and irregularities in the surfaces are leveled out. For packing, the package is simply worked over at forging temperature with a heavy hammer.

Borax powder is added as flux.

The package is heated to a yellow-red color, then welded with consistent hammer blows.

Drive the flux out of the gaps with hammer blows.

After the package has been welded, it can be forged out.

3.3 Folding the Package

To increase the number of layers, the welded package has to be folded. For this, it once again is heated to forging temperature (heat color: light cherry, about 900°C [1,652°F]) and stretched in length. Now, the round poll of the hammer pushes the material to the sides and thus stretches the workpiece.

After that, it is again heated to forging temperature, notched in the center with the forging chisel, then bent. The workpiece then is freed from forging scales by means of a wire brush, powdered with borax, once more heated up to an orange glow, and then bent together until only a small gap is left. The package is then heated up to forging temperature and welded together again with a few blows of the hammer.

It is advisable to remove the bending point using the anvil chisel—faulty welds often occur in this area.

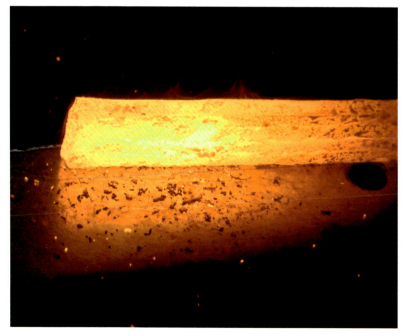

The package is repeatedly heated to forging temperature and beaten out.

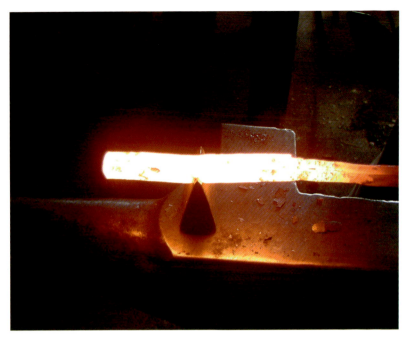

On the anvil chisel (a wedge) the package is notched at the center.

At the notch the package is bent over the edge of the anvil and welded together again.

The welded package is forged out again and stretched lengthwise.

The raw material is welded in the shape of a blade. The single layers are visible here.

SIMULATION WITH MODELLING CLAY

To explain the process a bit more clearly, we used modeling clay in various colors to reproduce forging a Damascus blade with an additional bar for the cutting edge.

Blue and yellow symbolize the steel for the side layers, red for the cutting edge.

The steels are stacked into packages and welded together.

The package is notched and folded.

The folded package is welded once again. The mixing of the colors blue and yellow in the border areas shows that the steels also get "mixed." Carbon moves from areas which are carbon-rich to areas which are poor in carbon.

A cutting layer is put into the center.

The "blade" is forged out.

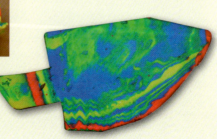

The number of layers is doubled with each folding so that with an initial package of five layers (in theory, after ten folds) more than 5,000 layers have been created. But this is just theoretical because outer layers continue to get burned up.

There is a difference of opinion when it comes to asking how many layers are the right number of layers. The simple formula "the more layers, the better" is by no means correct. The layers become ever finer and eventually can't be distinguished on the surface anymore. Besides that, the steels become more and more similar to each other because of carbon diffusion (from carbon-rich to carbon-poor layers). In addition, with each folding process there is the risk that welding won't succeed and gaps may form between layers. If this happens, as a rule, the workpiece is ready for the scrap iron container.

There are some Japanese swords with more than a million layers. But there is a specific reason for this: By folding, forging out, and welding steels with different carbon contents and impurities many times over, homogeneous steel is created. This means, the steel is mixed thoroughly like dough and impurities are removed in the process.

SOLVING A COMMON PROBLEM

What can you do when the weld of a Damascus package comes open? An old trick is to put the package into vinegar over night. The vinegar acts as acid and removes the oxide layer which otherwise would prevent successful re-welding. A welding package can't be rescued this way every time, but it is worth trying.

Problem package: the weld came open again.

CREATING PATTERNS

Early swords already distinguished themselves with beautifully patterned Damascus steel. These patterns, besides the decorative aspect, also had a practical function—the intense mixing of different steel types created a blade that was better suited for use in battle. In the following chapter we describe how different patterns can be created.

4.1 Layered Damascus

While welding the Damascus packages, without further work so-called layered or wild Damascus is made. By means of grinding the blade at the sides, the layers become visible. The pattern is random.

Layered Damascus.
Photo: Dick Feine Werkzeuge

Layered Damascus

Cut face of the layers when grinding the blade (schematic view)

4.2 Banded Damascus (Ladder Pattern Damascus)

Using different techniques patterns can be determined and created. The banded pattern is created by grinding grooves into the welded and forged-out piece of layered Damascus, i.e. by means of a disc grinder. Of course, the grooves can be filed as well, if there is no disc grinder at hand. In industrial production these grooves are made by means of a forging die.

After the grooves have been ground, the piece is forged flat again. Thus the deeper layers in the cut areas are brought up to the surface. A pattern is created which becomes visible later on during etching. If you grind grooves into the materials that are crossing each other, the pattern will become checkered or shaped like pyramids.

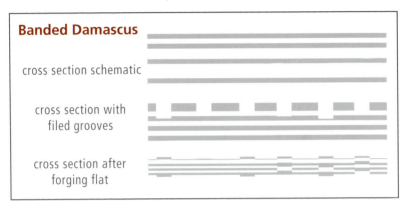

Banded Damascus

cross section schematic

cross section with filed grooves

cross section after forging flat

Grooves are created with the disc grinder.

Blade with banded Damascus (blade by Jürgen Rosinski, knife by Ernst G. Siebeneicher-Hellwig).

4.3 Rose Damascus

Rose Damascus, which looks like rose blossoms, is created by drilling blind holes in the piece of layered Damascus. The size of the rose pattern depends on the diameter of the drill bit. For our piece, the blind holes were drilled with a bit size of 8.0 mm (0.31 in). The depth of the holes is about five millimeters (0.196 in). The holes are drilled on both sides. After drilling, the workpiece is forged flat again.

For this, we chose a piece of layered Damascus with about 60 layers and about 12 x 35 x 100 mm (0.47 x 1.37 x 3.93 in) in size. Prior to drilling, the piece ought to be soft-annealed because otherwise drilling becomes strenuous. For soft-annealing, the workpiece is best brought to glowing temperature together with larger pieces of steel. Then it is left to cool in the forge overnight together with the other pieces. Cooling down should take place as slowly as possible. This is the reason for including the additional steel pieces which keep the warmth for a longer time.

The drill holes are first center-punched to provide a better grip for the drill bit.

Drill evenly spaced blind holes.

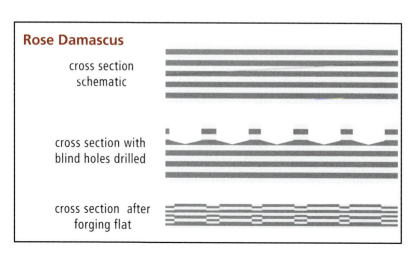

Rose Damascus

cross section schematic

cross section with blind holes drilled

cross section after forging flat

Damascus stamped with a casting die. After grinding, the pattern becomes visible.

"NAIL DAMASCUS"

An interesting variation for creating rose Damascus: Drill holes through a piece of layered Damascus then hammer nails into the holes. The whole thing is then welded in the fire (blade by Hans-Georg Ortner).

In the middle of the image, the "roses" are clearly visible on the ricasso. But where the blade begins, the pattern becomes indistinct because of being ground at an angle.

Blade with rose Damascus (knife by Richard Spitzl).

4.4 Twist Pattern Damascus

As with the banded and the rose Damascus, the starting material for the twist pattern Damascus is some layered Damascus. With both previous kinds of Damascus, the piece of Damascus was already forged relatively flat prior to further mechanical work, about down to twice the thickness of the planned blade. Here, the piece of layered Damascus should have a rather square cross section prior to twisting it. Its size depends on the planned size of the blade.

The workpiece is heated up to forging temperature, clamped into a vise, then twisted by means of tongs. From here you move on in sections. You can twist the whole rod in one direction, or twist the sections against each other, which leads to a different pattern later on. It is important to never twist back the already twisted area. Because then, with high probability, the welds for the single layers will come loose again, and the workpiece is ready for the trash can.

When making a twist pattern with Damascus, if you use non-hardenable steel then the layer for the cutting edge either has to be made of a hardenable monosteel or out of Damascus with only hardenable steels.

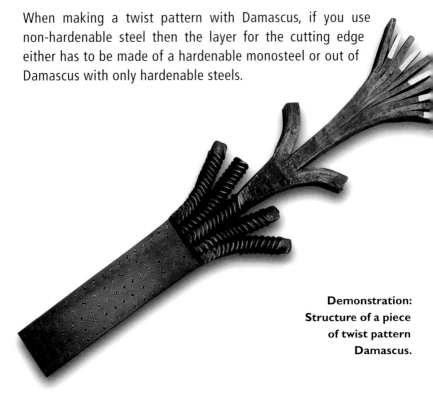

Demonstration: Structure of a piece of twist pattern Damascus.

The piece of Damascus is heated to forging temperature and twisted in the vise.

Here, the twists are clearly visible.

The twisted bar in the picture above was ground to different depths in several steps and was then etched to show the development of patterns at different levels.

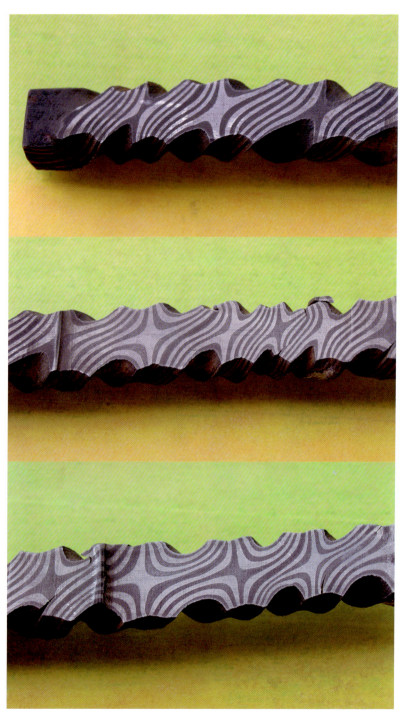

The various levels of the twisted bar shown in close-ups.

4.5 Multiple Bar Damascus with Cutting Area

An advanced way to forge Damascus is to weld together several bars and then sole on the cutting area. To achieve high contrasts, you can weld steels with low carbon content or pure nickel to carbon steels and twist them. Since this kind of twist pattern Damascus has "weaker," non-hardenable parts, a cutting area out of well-hardenable steel is forged on.

The package assembly prior to welding. Three bars of twist pattern Damascus (top) and a bar of layered Damascus made out of hardenable steels for the cutting area (bottom).

The layers are welded; the curve for the tip of the blade is already forged.

Blade made by Jürgen Rosinski out of twist pattern Damascus with a cutting area of layered Damascus. To enhance visibility, one of the three bars was dyed.

Blade of twist pattern Damascus with soled-on cutting area (by Hans Georg Ortner).

Blade with cutting area (by Manfred Ritzer).

Knife with blade of mosaic Damascus and cutting area of layered Damascus (by Luca Distler).

4.6 Mosaic Damascus

Compared to layered Damascus, to make mosaic Damascus, the pieces of steel are not simply stacked on top of each other but arranged in a three-dimensional pattern. For this, a multitude of pre-shaped elements often are combined to form complex structures. The welded pieces are forged out, cut, arranged into new packages, and welded again. Besides

To create a chessboard-like mosaic Damascus, square bars are stacked together to form a package. Both kinds of steel ought to have a high contrast after etching.

the problem of unreliable welding, the control of patterns is a challenge as well, because the pattern is distorted by the various working steps (welding, forging out).

The benefits of mosaic Damascus are of a strictly decorative nature—it doesn't offer any advantage with respect to the desired qualities of a blade. Thus mosaic Damascus is popular for bolsters and handle butts.

Knife with mountings made of mosaic Damascus (by Peter Huber).

While undergoing various processes, the pattern often gets distorted.

FORGING A BLADE

Out of the Damascus blank we now forge a knife with a hidden tang (on which the handle is mounted later on). For this, it is important that the steel doesn't get too cold. The heat color should not be below a dark cherry red.

We will demonstrate the individual steps to forging the blade using modeling clay—this makes it easier to see what is important. The starting material is in the shape of a round rod. Starting with flat material is even easier because you don't have to forge it flat.

Better for demonstrating: Our "round steel" is made of modeling clay.

The area of the blade is forged flat slightly at the beginning.

After that, the curve of the blade is forged (the workpiece is turned 90 degrees for this).

The area of the blade is widened further.

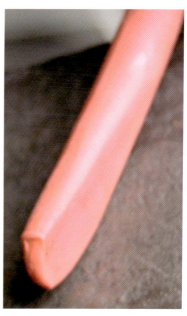

Towards the tip, the workpiece is reworked again and again between other steps.

The shape of the blade slowly reveals itself.

Towards the edge, the blade is forged flatter than at the back.

With the hammer peen, material is driven out to stretch the blade.

Traces of stretching are smoothed with the poll of the hammer.

The tang is established on the anvil's edge.

The area of
the tang is
smoothed.

With the hammer
peen the tang is
stretched in length.

This the tang is slowly brought to its final shape.

The result: our modeling clay knife, completely forged.

KNIFEMAKING

Even though the blade is already coarsely forged into shape, it has to be further worked on to bring it to shape. This is either done with a belt sander or manually with a file. We described how to grind blades in detail in *Basic Knifemaking: From Raw Steel to a Finished Stub Tang Knife.* Thus we don't elaborate on this process here. The blade is not sharpened at this point; this is done at the very end. We leave between 0.5 and 1.0 mm (0.019 and 0.039 in) of material for the cutting edge.

The process of heating and cooling the steel when forging has already made the steel relatively hard. Thus soft-annealing the piece is recommended prior to further work on the blade (see section 4.3).

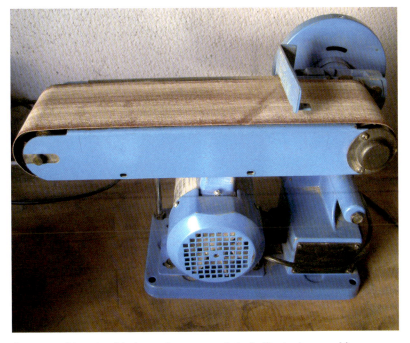

A reasonably priced belt sander saves a lot of effort when making a knife.

The blade receives its final shape on the belt sander.

6.1 Heat Treatment

Prior to hardening, all mechanical work steps, like grinding the blade and drilling the holes, have to be completed. During hardening, the steel is heated to hardening temperature. For carbon steels this temperature is about 900°C (1,652°F); at that heat the steel glows light red.

To determine the heat color correctly, the light inside the forge shouldn't be too bright. Also, when the steel reaches the right temperature for hardening, it loses its magnetic properties. This means a magnet can't attract the steel anymore. If you want to be absolutely sure you can use this method for checking.

Carbon steels are quenched in warm water or oil. We use vegetable oil (sunflower or rapeseed oil). For quenching, the blade is submerged tip first and slowly moved to and fro. To avoid scaling, protective lacquer has proven helpful. The blade is submerged in the lacquer and can be hardened after waiting for an hour.

WHAT HAPPENS DURING HARDENING

While being heated to hardening temperature, carbon atoms inside the steel move into "cubes" that are formed by iron atoms. When cooling proceeds slowly, then the carbon atoms move out of the center of these cubes again. During rapid quenching there is no time for this, so the carbon atoms remain "frozen" in position. The inner structure of the steel is thus distorted. This condition gives the steel its high hardness.

After quenching, however, the blade is much too hard. To achieve a good hardness for use, part of the hardness has to be taken out of the steel again. This is achieved by once again heating the steel. Annealing carbon steels is done from 100°C (212°F) up to 300°C (572°F). The proper duration for annealing is about one to two hours. You can determine the temperature of the steel by means of the colors created.

WHAT HAPPENS DURING ANNEALING

During annealing, part of the carbon atoms move back to their original place. The steel relaxes again. Its hardness, which was too high for use, reverts to a usable hardness. The hardness of knife blades is measured in Rockwell grades on the C-scale (HRC). Utility knives, like butcher knives or European kitchen knives, mostly rank in the area of 54 to 56 HRC; high-quality hunting knives rank between 58 and 60 HRC. The cutting layers of Japanese kitchen knives achieve a hardness of up to 66 HRC.

6.2 Etching

Etching the blade allows the pattern to be visible. It would be a pity if a pattern created with a lot of effort was only slightly visible or not at all. Depending on the carbon content, the etched image is darker or lighter in color. Layers of St37, with its low carbon content, stay bright during etching. Nickel stays very bright, while steels with a high amount of manganese become dark.

For etching you can use hydrochloric acid (available at the pharmacy), sulfuric acid (can be bought at the gas station as battery acid), or ferric chloride (used for etching circuit boards and usually available at electronics shops). Ferric chloride is the least dangerous, it doesn't burn the skin but does stain yellow if it is spilled.

The depth of the etching is dependent on three factors: the concentration of the acid, its temperature, and the duration of the etching process. Deep etchings, which result in relief-like structures, are best created with warm sulfuric acid at 35%.

Prior to etching, remove any oils or fats from the blade with acetone. The acid is then put into a tall glass. The procedure is easy: heat water in an old pot, set the glass with the acid inside the water bath, wait until the acid is warm, then hang the knife inside (prior to this, heat the glass with warm water, so it doesn't crack due to the difference in temperature).

After a short time bubbles rise up, and soon foam will form on the surface. After about five minutes (with ferric chloride) the blade should be taken out of the bath and the depth of the etching should be checked. For this you can put the blade into a receptacle with water and rub it with an (old) kitchen sponge.

If the depth of the etching or grade of coloration is sufficient, the blade is rinsed in clear water and neutralized with soda (baking powder) so the acid will stop working. For neutralizing, you can put the blade onto a flat plate and strew the baking powder over it. Then pour some water over it and let it take effect for a couple of minutes. The dry blade can then be polished. If the blade is only polished lightly, the pattern stays gray and rich in contrast. If polished more intensely, the gray color vanishes and only the relief is left behind.

 **IMPORTANT:** Etching should always be done outdoors; do not inhale any vapors throughout the process; and wear protective goggles and gloves!

The acid bottle in a water bath. From the pencil dangles the yarn which holds the knife.

A Damascus that is rich in contrast from the combination of structural steel St37 and 1.2510 (O1) tool steel.

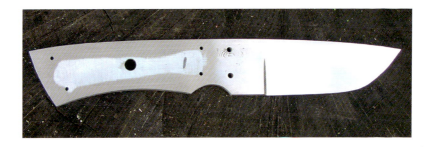

A Damascus blade prior to etching…

… and after etching and polishing.
Only the etching lets the pattern become visible.

Deeply etched Damascus (knife by Ernst G. Siebeneicher-Hellwig, handle of fossilized wood).

6.3 Constructing the Handle

Since we built a hidden tang knife in *Basic Knifemaking: From Raw Steel to a Finished Stub Tang Knife*, here we want to introduce a full tang knife. This means, our knife will have two handle scales which are mounted to the left and right onto the tang. Between these scales and the blade we want to attach bolsters.

Before we start with the construction of the handle, we wrap the blade with adhesive tape in such a way that it is completely covered up to the handle. This will prevent the blade from being damaged while working on the handle.

The bolsters are sawed from a piece of nickel silver and are pre-shaped. Finish and polish the bolsters because it is hard to polish this area later, when the bolsters are already riveted.

Drill holes for the rivets into the nickel silver. When drilling through, burrs are created on the underside which have to be removed. For this we use a flat file. The areas which are resting against the steel have to be ground absolutely level to avoid a visible gap between blade and bolsters. This is best done on a flat surface.

Pre-finished bolsters with holes for the rivet pins.

After the burrs have been removed, the rivets are put into the pre-drilled holes of the knife, the bolsters are put on, and the whole assembly is riveted.

One strategy that has proven useful is to slowly press the rivets together in the vise prior to riveting with the hammer. For this, the ends of the rivets (which ought to project about two millimeters [0.078 in] on each side) are clamped between the jaws of the vise and pressed together with the material of the bolsters.

The advantage of this process is that the riveting is done in a controlled way, and possible slipping can be corrected.

First, the rivets are pressed together in the vise in a controlled way.

Finally, the rivet heads are pounded flat with the hammer on an anvil, vise, or a sheet of iron.

For this we hold the knife in such a way that only the ends of the pins, which project from the bolsters, lie on the hard base. Then we make slight hammer blows on the opposite end of the pin. By doing this, the projecting end is pressed flat.

The rivet ends are then hammered flat on an anvil.

HANDLE MATERIALS

- nickel silver, flat material, 7 mm (0.28 in) thick
- tube for lanyard hole
- nickel silver pins
- material for the handle scales

TOOLS AND ACESSORIES

- vise
- abrasive cloth, grits coarse, medium, fine
- hacksaw
- flat file
- hammer
- semicircular file
- abrasive rolls
- epoxy
- key files
- drill press
- drill bit for wood
- polishing set

To seal the space between bolsters and tang, you can apply superglue prior to riveting.

We prepare the handle material. It is easiest if you obtain prepared handle scales from a specialized dealer. But of course, you can saw appropriate pieces of wood, horn, or bone yourself. The material has to be absolutely level on its underside, where it rests against the tang. As with the bolsters, it is recommended that you grind the flat area level on a flat surface.

On the front side, where they rest against the bolsters, both handle scales are filed or ground at a right angle to the supporting area. The pieces are put onto the tang, and the contour of the tang is transferred onto the handle material with a pen. Afterwards the handle scales are sawed (leave enough excess for finishing later).

Prior to gluing the scales, we drill a row of blind holes into their underside. For this we use a drill bit for wood or steel with an approximate diameter of 5.0 mm (0.196 in); the size isn't a crucial detail. The depressions serve as reservoirs for the glue. When the flat tang is fitted to the flat handle scales, the glue can get out of the way and move into the depressions, so it won't be squeezed to the outside.

Prior to gluing, remove any fats or oils from the tang with acetone. Rough the surface of the tang up with a coarse sandpaper. Mix the epoxy as described by the manufacturer and apply it on one side of the tang. Then press one handle scale on and fix it with clamps.

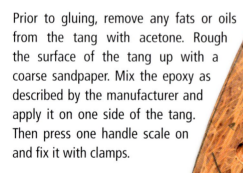

Blind holes in the handle scales serve as glue reservoirs and take in the glue later on.

One handle scale is already glued and fixed with clamps.

After the glue has hardened, the holes for rivets and lanyard are drilled into the wood.

The holes already drilled into the first handle scale act as a guide for the holes that go through the second handle scale (lower down in the image).

After the glue has hardened, remove the clamps and drill the holes for the rivets and the lanyard. The holes that are already drilled in the tang act as guides.

Then glue the second handle piece on. After hardening, the holes for pins and lanyard are also drilled through the second handle scale. For this, holes that were already drilled for the first handle scale act as a guide.

The diameter of the drill holes ought to be a tenth of a millimeter (0.003 in) larger than the diameter of the rivet pins and the tube for the lanyard. Since the pins and the tube are glued in later on, it isn't necessary to press the parts firmly into the material. The pressure created by this pressing could lead to cracks while riveting, or if the material shrinks later on.

The pins are put in place. They are glued, but not yet fastened with hammer blows.

Rivets and lanyard tube should be roughened with sandpaper and all oils should be removed with alcohol or acetone. After this, both are painted with glue and pressed in. It is not recommended to rivet the handle scales with hammer blows.

Now the handle scales are worked on with files and sandpaper until they have the desired shape. The sandpaper is used to remove the unwanted file marks. While working, change to the next finer file cut when the traces of the previous work step are no longer visible. After filing, remove the traces of the files with the coarse sandpaper. Only when all traces of the files have been removed do you use the next, finer grit. In general, switch the direction of sanding with each change of grit by 90 degrees.

When working with sandpaper, it is helpful to tear the paper into small strips and put one across the blade of a file. Then you can hold it with a finger and guide the file with the sandpaper over the workpiece. Abrasive rolls, clamped in the drill press, are well suited for creating radii at the handle.

Finally, our handle receives an oil finish: It is painted with cooked linseed oil mixed with turpentine. After a couple of days for drying, the wood is polished with fine steel wool and painted once more. We repeat the whole procedure a couple more times. Finally, the handle is polished manually with an old woolen sock.

A polishing set from the home improvement store for working on the handle.

The finished knife with polished handle scales.

Abrasive rolls for the drill press are helpful for working out radii (at the bolsters).

6.4 Edge Sharpening

An important finishing step is sharpening the blade. There are many methods and opinions regarding this process. Hopefully it is now common knowledge that sharpening blades on fast-running machines without cooling the steel is not good for the blade. The heat generated by the friction in this crude action takes the hardness out of the blade.

Only prior to heat treatment is the use of machines, such as the belt sander, possible. The unwanted effects of too much heat only show up after hardening and annealing.

There is a broad range of sharpening accessories and tools available which indeed have their advantages. In our opinion, sharpening on Japanese waterstones is the optimum method. According to our own experience, the way Japanese samurai have achieved legendary sharpness on their weapons since the early Middle Ages is the best method, even nowadays.

Certainly, at the beginning it takes some training to achieve the desired result. But if somebody takes the effort of forging a blade and making a knife from this, then learning how to sharpen it will be the least problem.

With Japanese waterstones, wet sharpening is the general rule. For this, prior to sharpening, the stone is put into a bath of water for about ten minutes. The stone is soaked when bubbles no longer rise. The water on the stone provides cooling and is also responsible for the removal of metal parts and the rubbed-off parts of the stone. For this, every once in a while some water should be poured onto the stone during sharpening.

Japanese waterstones are available in grits from very coarse (80) to very fine (16,000). Two-layered stones are also available; they combine a sharpening stone and a stropping stone, i.e. a 1,000/6,000 combination. For preliminary grinding, a coarse sharpening stone of grit 600 is suitable.

A Japanese waterstone clamped in a very practical fixture.

The stone is then put onto a skid-proof base on top of the table. Fixtures for sharpening stones, which are available at stores, provide a secure base, so the stone can't skid.

The blade is put onto the stone diagonally, at an angle of about 45°. It is tilted in such a way that the desired edge angle will be created. For kitchen knives, a flat angle is chosen (10° to 15°). For a more robust hunting knife, which is also used for working with wood, the angle should be steeper (20°). It is important, that the angle is evenly maintained during sharpening. Tipping movements create a round edge which isn't sharp.

The handle rests in the right hand while the blade is pressed onto the stone with the fingers of the left hand (for right-handers). For sharpening, the blade is drawn in a straight line over the length of the entire stone. During both movements—pushing as well as pulling—slight pressure is put onto the blade. The abrasion of the blade can be seen as a dark track on the stone. After pushing and pulling for about 20 to 30 times, the resulting sharpness is checked by lightly sliding a finger tip across the edge from the non-sharpened side. If a palpable burr has been created, then you can sharpen the other side. Now, the knife is simply turned around in the

The correct handling of the knife during sharpening.

The sharpening angle (here a flat angle for a kitchen knife) has to stay constant.

hand. The other side is sharpened until a tangible burr has been created there as well.

During the subsequent stropping, the knife is guided over the fine stone the same way as before. The pressure against the stone should be a bit less, since the blade edge now only has to be polished and the last, fine burr has to be removed. Besides that, the finer stropping stones are softer, too, and can be damaged by intense pressure.

Sharpening the opposite side.

The burr created when sharpening is easily felt.

6.5 Testing the Sharpness

The most common test with respect to blade sharpness is the razor test on the forearm. But: What's a guy going to do without dense hair on his forearm? One alternative is the paper test. For this, a sheet of paper which is held freely is cut into pieces. The blade should glide through the paper almost without resistance and without tearing.

You can also do the tomato test: Put the tomato onto the cutting board, hold it, put the blade on it and draw it without pressure across the tomato. The blade should literally sink into the tomato under its own weight only.

The paper test is an alternative to the razor test.
Photo: Manfred Ritzer

For the tomato test, the blade is guided without pressure.

DMT SHARPENING SYSTEM

For beginners, practical sharpening systems, like the one from the company DMT (US), are available. Even those who are just starting out will enjoy rapid success with this option. The device works using two diamond sharpeners (with two different grits) and a clamp which holds the blade. The sharpening angle can be determined and the blade holder can also be used for guidance, for example, when sharpening on waterstones.

A fixture is clamped to the blade. The angle can be adjusted.

The sharpening element is held at a constant angle by means of a guidance device.

The DMT clamp can also be used for sharpening on a stone.

6.6 Protection against Rust

Since only steel types that are not stainless are suited for conventional welding in the forging fire, protection against corrosion of the knife is an important topic. It is important to clean the knife after use with water and to dry it well afterwards. A few drops of acid-free oil (camellia oil for example) are a treat for the blade, especially if it has not been used for a long period of time.

A well-proven method against rust is black annealing. This process was first applied way back during the Middle Ages to protect iron parts against rust. The layer of burnt oil which is created by quenching in oil is simply left on the blade. It protects against rust like a coating. A similar way is it to leave the hard layer of cinder created during forging (called forging scales) on the blade. This layer also reduces the risk of rust.

Unfortunately, with both methods the patterns on a Damascus blade can't be seen anymore. Thus this method only makes sense for sandwich blades with a central cutting layer (as are common with many Japanese knives).

Multi-layered blade with forging scales in the upper area.

Sandwich blade with forging scales and clearly visible traces of the hammer.

APPENDIX

7.1 Heat Colors for Forging and Annealing

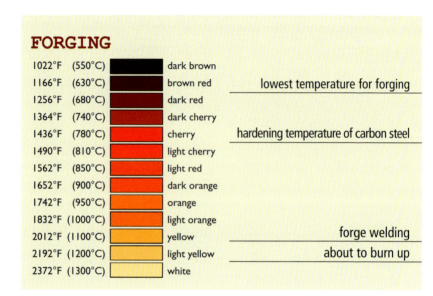

FORGING

1022°F	(550°C)	dark brown	
1166°F	(630°C)	brown red	lowest temperature for forging
1256°F	(680°C)	dark red	
1364°F	(740°C)	dark cherry	
1436°F	(780°C)	cherry	hardening temperature of carbon steel
1490°F	(810°C)	light cherry	
1562°F	(850°C)	light red	
1652°F	(900°C)	dark orange	
1742°F	(950°C)	orange	
1832°F	(1000°C)	light orange	
2012°F	(1100°C)	yellow	forge welding
2192°F	(1200°C)	light yellow	about to burn up
2372°F	(1300°C)	white	

ANNEALING

392°F	(200°C)	bright metalic
410°F	(210°C)	yellowish white
428°F	(220°C)	light yellow, straw
446°F	(230°C)	yellow
464°F	(240°C)	deep straw
482°F	(250°C)	yellowish brown
500°F	(260°C)	reddish brown
518°F	(270°C)	red, purple red
536°F	(280°C)	violet
554°F	(290°C)	dark blue
572°F	(300°C)	cornflower blue
590°F	(310°C)	light blue
608°F	(320°C)	greyish blue
626°F	(330°C)	grey, greenish grey

Depending on the annealing temperature and the type of steel, the workpiece takes on different colors. In general, the higher the annealing temperature, the darker the color. The spectrum ranges from yellowish white over orange and red hues to blue and grayish green.

7.2 Recommendations for Safety

Observing the following recommendations should go without saying, but doctors and nurses can tell you a thing or two about how careless some craftsmen are when dealing with their health and safety.

- Follow the manufacturer's instructions and guidelines when working with machines.

- Wear protective goggles.

- Wear ear protection.

- Don't wear loose clothing or jewelry while working with machines with rotating parts.

- Solvents and acids are to be used outdoors only. Wear protective gloves and goggles while working with them.

- Wear dust masks or respirators during all kinds of work where dust is created. The fine dust created while grinding can cause severe and lasting damage to your respiratory system.

- While forging: always wear protective goggles, wear solid shoes, and solid clothing (not made of synthetic fibers!). Protect your hands with thick gloves.

- While working on the tang and handle, cover the blade with adhesive tape.

Always follow the safety recommendations and rules when working with gas!

7.3 List of Suppliers

Alpha Knife Supply
(425) 868-5880
chuck@alphaknifesupply.com
www.alphaknifesupply.com

Culpepper & Co.
(828) 524-6842
info@culpepperco.com
www.knifehandles.com

Damasteel AB (Sweden)
(46) 0-293-30600
mail@damasteel.se
www.damasteel.com

Halpern Titanium
(888) 283-8627
info@halperntitanium.com
www.halperntitanium.com/knife.htm

Hawkins Knife Making Supplies
(770) 964-1023
sales@hawkinsknifemakingsupplies.com
www.hawkinsknifemakingsupplies.com

Jantz Supply
(800) 351-8900
jantz@jantzusa.com
www.knifemaking.com

Knife and Gun Finishing Supplies
(800) 972-1192
csinfo@knifeandgun.com
www.knifeandgun.com

NorthCoast Knives
pjp@NorthCoastKnives.com
www.northcoastknives.com

Provision Forge
(541) 846-6755
bladesmith@customknife.com
http://customknife.com

Texas Knifemaker's Supply
(888) 461-8632
sales@texasknife.com
www.texasknife.com

Thompson's Scandinavian Knife Supply
(517) 627-9289
bradjarvis3@comcast.net
www.thompsonknives.com

Tru-Grit Inc. (Canada)
(909) 923-4116
www.trugrit.com

USA Knife Maker Supply
(507) 720-6063
info.usakms@gmail.com
www.usaknifemaker.com

7.4 Where to Learn Forging

Contact these workshops for the most up to date information before making arrangements to participate.

Bighorn Forge Ironworks
(262) 626-2208
4190 Badger Rd.
Kewaskum, WI
www.bighornforge.com

New England School of Metalwork
(888) 753-7502
7 Albiston Way
Auburn, ME 04210
www.newenglandschoolofmetalwork.com

North Bay Forge
(360) 317-8896
1 North Bay Street
Waldron Island, WA 98297
www.northbayforge.com

Pouncing Rain Jewelry and Metal Working Center
(360) 715-3005
521 Kentucky Street
Bellingham, WA 98225

David Robertson
Artist Blacksmith
(519) 366-2334
R.R. #2
Cargill, Ontario,
Canada, N0G1J0
www.artistblacksmith.com

R. Twerks (UK)
Nick Johnson
(+44) 1553-810846
www.knivesbynick.co.uk

Spitfire Forge
Women's Welding Workshop
P.O. Box 658
El Prado, NM 87529
www.spitfireforge.com

University of Kentucky
Department of Art
(859) 257-8151
207 Fine Arts Building
Lexington, KY 40506
www.uky.edu/FineArts

D.E. Wilson Forge
(207) 348-6871
455 Eggemoggin Rd.
Little Deer Isle, ME 04650
www.dewilsonforge.com

7.5 A Meal for Bladesmiths

During a gathering of bladesmiths at the workshop in Thaldorf, Germany, a nourishing meal was prepared. As with our first book, we want to share the recipe with our readers here, too.

LIST OF MATERIALS

- 5 pound of pork shoulder, cut
- 4 onions, cut into quarters
- 6 carrots
- 2 cloves of garlic
- 6 potatoes, sliced
- 1 slice of celery
- ginger
- pepper and salt
- 100 grams (1/2 cup) of lard
- dark beer

Doesn't this look delicious?

DIRECTIONS:

Put the lard into a wrought iron roasting pan. Put the shoulder piece skin down in the pan together with the other ingredients. Roast for one and a half hours at 180°C (~350°F).

Stir the ingredients, flip the shoulder over so the skin is on top, and stew covered for another one and a half hours. Turn the heat up to 220°C (~425°F) and remove the lid until the skin caramelizes. Pour dark beer over the shoulder every once in a while.

Bon appétit!   *Recipe: Hubert Ziegler*

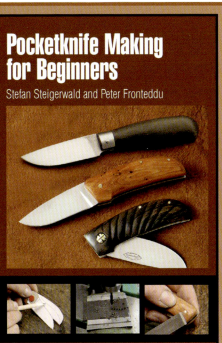

Pocketknife Making for Beginners.
Stefan Steigerwald & Peter Fronteddu.
Make your own folding pocketknife with this easy-to-follow guide. Step by step, this instructional manual unfolds the secrets of constructing a slip-joint folding knife, which is held open by spring force and friction. In addition to introducing different variations of this knife style, this guide presents the materials, tools, and technical design skills needed for the project. Diagrams clearly demonstrate the mechanics of your knife and the crucial elements needed to make a properly functioning pocketknife. Detailed step-by-step explanations move from template to finished knife — even beginners can master this project with minimal tool requirements. Once the knife project is complete, you can use the processes in this guide and your own creativity to construct a special knife of your own design.

Size: 6" x 9" • 275+ photos & diagrams • 128 pp.
ISBN: 978-0-7643-3847-2 • spiral bound • $29.99

Basic Knife Making:
From Raw Steel to a Finished Stub Tang Knife.
Ernst G. Siebeneicher-Hellwig and Jürgen Rosinski.

In this book Ernst G. Siebeneicher-Hellwig and Jürgen Rosinski show the simplest and least expensive ways to construct a simple forge, make all necessary tools yourself, forge a stub tang blade from an old automobile coil spring, and make a complete knife.

Their practical guide demonstrates the most important theoretical basics and shows how simple it can be to experience bladesmithing. Each step is presented in text and pictures, with a special focus on forging the blade. Clear lists of tools and materials help you through the process. Practical tips, explanations of terms, and sketches round out the volume.

Size: 8 1/2" x 11" • 205 color images/10 drawings • 112 pp.
ISBN: 978-0-7643-3508-2 • soft cover • $29.99

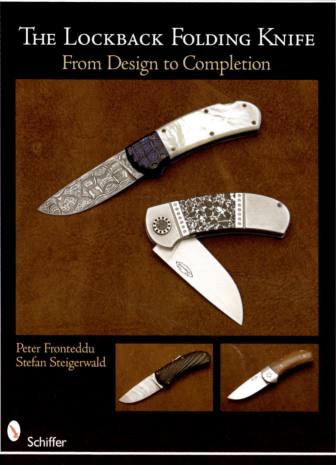

THE LOCKBACK FOLDING KNIFE
From Design to Completion

Peter Fronteddu
Stefan Steigerwald

Schiffer

The Lockback Folding Knife: From Design to Completion.
Peter Fronteddu and Stefan Steigerwald. Take your knifemaking
skills to the next level and create your own folding lockback
knife. Illustrated instructions and more than 200 color images
detail all stages of the knife's construction, from creating a
template to making the blade and locking mechanism. With
this guide you can tackle the challenging task of constructing a
lockback knife and gain the skills necessary to create a lockback
knife of your own design.

Size: 8 1/2" x 11" • 236 color images • 112 pp.
ISBN: 978-0-7643-3509-9 • soft cover • $29.99

Making Integral Knives

Peter Fronteddu and Stefan Steigerwald

Making Integral Knives.
Peter Fronteddu & Stefan Steigerwald.
The next installment to this knife workshop series explains how to design and build an integral knife, a knife made out of a single piece of steel. From basic patterns and principles to technical solutions to various variations in design and process, this guide is ideal for the intermediate to advanced knifemaker. Through step-by-step instructions and images, three integral knife projects with varying levels of difficulty are explained. Learn how to make a hand-filed knife that doesn't require much equipment, a knife made using a milling machine, and a knife made with a piece of steel that has been professionally prepared with a wire-erosion process. With 350 photos and illustrations, this comprehensive guide is ideal for mastering how to make integral knives.

Size: 6"x 9" • 350 photos & diagrams • 144 pp.
ISBN: 978-0-7643-4011-6 • spiral bound • $24.99